gENESIS
FROM NOTHING TO NOW, AND BEYOND.
"A 'Briefer' History of Time"[1]
By
Robert A. Tallarico
Copyright 2017 by Robert A. Tallarico

Front cover design and painting "gENESIS"
By
Robert A. Tallarico

Also by

Robert A. Tallarico

A MUSING
Life's Big Little Stories

Oh My GOD !!!
From Doubter to Believer
Via Science and Theology

Available on Amazon.com
Books
Robert A. Tallarico

And Other Retailers

Work In Progress

MANGIA! MANGIA!
Per Favore.
Recipes for Family Cooking

INTRODUCTION

This booklet was written to clarify, consolidate and simplify the many sources of knowledge available today which when taken together leads to too much information and results in confusion and disinterest for many including some who practice in the fields of science, technology and theology.

This effort then is meant to provide a framework upon which the many areas of science, theology and philosophy may be inserted to broaden specific areas for a more detailed picture of the process from the void to now and beyond.

The Epilogue "End of Life" discusses a subject that surfaces for us all, especially those of us who have attained the wisdom of age. It ties into thoughts contained in the previous chapters of "gENESIS, From Nothing to Now, and Beyond.

I recently visited my Personal Care Physician with a list of the burdens that come with growing older. I learned, not to my surprise, that there is no cure for old age.
This encounter led me to thinking seriously about the ultimate climax to living. Not a

new effort on my part. Having been raised in the theology of Catholicism a lot of the time in contemplation is spent thinking about it. It is, however, a climax that I now take more seriously. Seriously enough to consolidate my thoughts and write about it.

It is my hope that others who may read this booklet will share their thoughts with me so that I and they may develop a more complete if not perfect idea of the happening.

The cover painting genesis is a contemporary take on the Adam and Eve story which appears in the Bible. It deviates from the biblical story and tries to present a more acceptable understanding based on modern knowledge. It should be more acceptable to the younger generation schooled more heavily in science and technology.

It is hoped that this interpretation is not offensive to those without the scientific revelations of the day or who instead prefer not to delve too deeply into the biblical story.

Robert A. Tallarico
July 2017

CONTENTS

FROM NOTHING

The problem of how visual light and other electromagnetic waves travel through the void of space, that is through a medium of nothing, has recently been resolved by the genius of science. From the time of Maxwell's brilliant science changing formulations of electro-magnetism resulting from the interaction of wires carrying an electric current, and magnets surrounded by a magnetic field, to Einstein's thoughts that this phenomena occurs in the void of space without magnets and wires, or any physical medium, the possibility exists that prior to the big bang there existed nothing other than a potential field for something. A field used here is something intangible like a gravitational or magnetic field.

This may sound like mumbo-jumbo but it is worth considering from a philosophical/theological point of view in spite of the physicists and cosmologists of our day abhorring the adulteration of science and mathematics with non-provable theories of philosophy and religion. This point of view requires us, as imperfect intelligent beings, not to look back in time to creation, i.e. to "The Big Bang," and the creation of Stuff but instead to look forward from the period before the act of creation of "Stuff." A period of nothing.

To do this we must first of all get rid of the physical part of us, that is our bodies, and think of ourselves as pure mind just as the electromagnetic phenomena gets rid of wires and magnets and is a purely electro-magnetic function or field. With our mind only, construct the nothingness of the void in an overly simplified way by imagining an infinite number of sinusoidal waves of equal numbers in and out of phase. Attachment (1). This simplification will be useful in conveying the idea. Now label half of the waves positive, and the other half negative. Or imagine a +1 and a -1. As these waves or numbers combine they cancel each other out such that they equal a zero amount of Stuff or nothing.

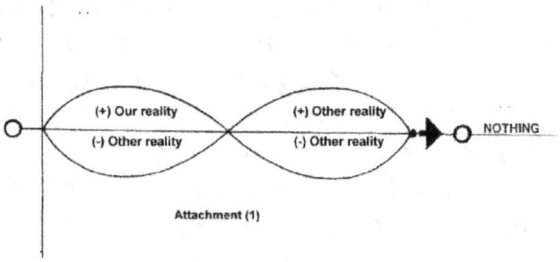

Attachment (1)

Now imagine a superior intellect without a physical presence, somewhat like us without bodies. A superior intellect that in my opinion is an infinite genius. This genius can easily turn in it's mind our mathematics or whatever superior function is necessary to initiate the creative process of separation of nothing into something. It does this by force of mind.

This genius, for whatever reason, by will of mind, out of boredom with the void or an itch to create and play with mathematics, energetically separates the void into positive and negative elements. The positive, which adheres to our current thinking, suddenly becomes the moment of creation of our universe. Dramatically called "The Big Bang," it results in Stuff (mostly hydrogen) and ultimately us. The other half unseen by us does something similar beyond our ability to perceive its construct. Current thinking is that it is identical to our reality except with an opposite charge. It is known that when an electron, our world, collides with an opposite charged positron, the other world, that the two annihilate each other, release energy, and result in zero stuff.

Taking the idea further, what we consider our reality occurred over a vast area of void instantaneously. Our view of it is limited in dimension by the maximum speed of light. The actual size of creation may be an infinite Macro Universe of which our reality which is visible to us is one small piece

TO NOW

Fourteen billion years is a long time. To put it in perspective think fourteen thousand million years. That is the approximate amount of time from nothing and the first instant of creation to now. It is the amount of time it has taken creation's light to reach us. It is the approximate age of our Universe.

To sense the size of the universe, multiply that number of years by the speed of light, commonly noted as "c", which is 186,000 miles per second.[2] This calculation, correcting seconds to years, gives us the size of the Universe in miles which is truly astronomical. The calculation ignores the expansion of the universe which has occurred over this same period of time making the size of our universe even bigger.

14 billion years x c =
82,120,000,000,000,000,000,000 miles =
8.212×10^{22} miles.

———————

From the first moment of creation until about seven billion years ago, the first atoms that existed, hydrogen and some helium[3] were subjected to gravity, fused and formed the first stars and galaxies which now are no longer in existence. About this time thru the attractive force of gravity on the materials in the universe and the explosion of giant stars into supernova, the elements that exist in the universe today were created and are the components of all universal material and life. Prior to this time, based on current understanding, there was no possibility for life as we know it because the materials necessary to form life did not exist in quantity.

From seven until about five billion years ago, the materials necessary to create our solar system, and probably other systems throughout the universe, accumulated into a large sphere of gas and a sprinkling of heavier elements, and as it rotated it began to flatten into a disc. Through gravitational attraction, the major portion of the gas in the disc coalesced into a sphere in its center with a ring of heavy and light gaseous material around it. The sphere of gas, mostly hydrogen, was compressed by gravity and became very dense and hot. This condition caused the hydrogen to fuse together into helium. Radiation which we call sunlight resulted from this nuclear fusion and thus five billion years ago our star, the sun, was born.

The disc, or ring, surrounding the sun coalesced into the large gas planets that exist today in the outer reaches of the solar system. Nearer to the sun the rocky planets, of which the earth is one, also coalesced. Our planet earth is about four and a half billion years old. The asteroid belt which exists today is a ring of left over material that has not yet coalesced into a major planetary body.

Our planet earth, four billion years ago, was a hot, fiery and explosive ball bombarded by asteroids and comets which brought differing types of molecular material from outer space. From this initial start it went thru various stages of physical development including the creation of the moon which in theory, debatable, resulted from the impact of a small planet sized body and the ejection of planet and earth debris into space. This ejected material ultimately coalesced into the moon.

Earth ultimately cooled down enough to where a solid planet formed containing water and molecules, such as amino acids and nucleotides necessary for building life. The origin of these molecules came possibly from outer space which seems to have an abundance of them or possibly from ocean vents where they are generated chemically. Or possibly from both. Once available these molecules were able to form chemically into RNA with the ability to replicate. This was followed by the combination of RNA which was able to produce proteins. Thus followed the beginning of biology. LIFE. The proteins produced had no special order. They were randomly coded and as such produced only mutations. To organize the assembled molecules so that a structure could repeat itself with no or very few mutations DNA, which is similar to RNA, was assembled within a cellular structure and contained a fixed code for reproduction. This is the fixed arrangement or code written with four molecules and structured into a helix. When split longitudinally, each half of the helix combines with free molecules to recreate itself.

DNA however by itself does not produce proteins. This function is still unique to RNA which takes the code from DNA and uses it to produce proteins. In this way one celled living organisms started a cycle of living things about three billion years ago.

About five hundred and fifty million years ago, a period know as the Cambrian, one celled organisms in mud and water joined together to form two celled and ultimately multi celled life forms that worked together symbiotically to survive and reproduce themselves. This was the start of life in its simplest form.

Ultimately these multi-celled life forms joined in symbiotic relationships with other multi-celled life forms and became increasingly complex. The oceans were alive, and vegetation in the oceans and on land could now convert the energetic radiation from the sun thru photosynthesis into usable energy and thus grow into vegetative reproductive life. The earth turned green.[4]

Insects evolved on the land, and in the oceans animals thrived. Some with primitive lungs were able to gulp atmospheric oxygen and were finally able to develop lungs and exist on the land out of water. Life was good for these animals for a period of time because there were no predators on the land and there was an abundance of food in the form of vegetation and insects. Flowers evolved and produced seed, a concentrated form of energy, to be eaten and distributed by the animals and wind. This Utopia did not last long however because other animals evolved that preyed on the vegetarian and seed eating animals and on each other.

This condition existed over a period of about five hundred million years. Halfway thru this period, about two hundred million years ago the planet had one super continent called Pangaea. Life forms wandered over this continent and faced several mass extinctions leaving fossils which we can now study and use to establish the existence of this super continent and the evolution of life on earth. During the period from the Cambrian thru the late Triassic, life experienced and survived four mass extinctions. Innumerable forms of life evolved, adapted and perished. After the fourth extinction life on the planet was dominated by dinosaurs and lizards large and small.

About sixty five million years ago, which was the end of the cretaceous period, the planet experienced the fifth and last major extinction when many volcanoes erupted and a huge asteroid crashed into the earth causing enough debris for a long enough period of time to fill the atmosphere with dust and debris so that the energy from the sun could not penetrate sufficiently to energize plant life. This caused plants and plant eating animals to starve and perish, which in turn caused the carnivorous animals to go hungry and perish as well. It was the end of the dinosaurs on earth.

Along with the dinosaurs about 75 percent of life on earth perished. However avian dinosaurs and some mammals including a small mammalian rodent sized animal survived this extinction. This small animal evolved over the last sixty five million years to the present to become most of the animals (birds excepted) that exist on the earth today. One of which is us, the human species, a branch of hominids.[5] Current thinking has our species, homo sapiens, originating some 200,000 years ago in Africa. From there around 100,000 or so years ago the species migrated out of Africa toward the Middle East (Hello Adam and Eve) and then throughout the world. It is enlightening to know that we are mammals, and can only wonder what form intelligent life would have taken if the dinosaurs had not perished but continued to evolve in intelligence and skill to the present time. Females would likely produce eggs outside their bodies like the birds, reptiles and lizards. Probably nest with them until they hatched at which time the little creatures would, after a short period, fend for themselves. It might be a simpler world, and perhaps a better world, if we were all chickens. Imagine no women with breasts to make milk to feed their young. And no men to lust after, fight over and long for the love and attention of women. And vice versa of course.

So where does this dissertation lead?

With the human species now totally dominant over the earth. And with the species numbers increasing at what is almost a vertical unsustainable curve. And with evolution still a force of nature, probably more dynamic today than in most of the history of the planet because the change to the environment is occurring so rapidly that the natural forces do not have time to adjust. The fouling of the atmosphere by man's mining and returning the carbon, which was taken out of the atmosphere billions of years ago, buried and turned into coal and petroleum, back into the atmosphere, diluting the percentage of oxygen like existed when the earth was green without animal life. The warming of the oceans and it's subsequent rise and turning acidic by absorbing atmospheric carbon destroying the nurseries of the oceans, its coral reefs. And most importantly the knowledge of the condition but the greed and lack of motivation to adjust by the dominant animal which has the intelligence but not the wisdom to change. Another extinction following a short Anthropocene, a sixth major one, has started, and seems inevitable. This is not the ending of this section that I intended, but is the ending that weighs heavily on my mind and will not be denied.

AND BEYOND

What fun to project our knowledge into the future. Will the universe end, and how will it end if it ever does? Will it end in fire or ice or will it go on expanding forever?

The question although a challenge is really moot. The reason being that we fail to consider our own evolvement which will lead to a state of mind that goes way beyond our current physical and mental existence.

I would speculate that the future of the universe will be much like the "From Nothing" period. It will expand to where its density approaches zero, and there is nothing in existence but the genius mind field and the potential for the void once again to become Stuff.

How can this be? Consider, as is commonly done, an analogous state of the universe as a balloon with dots located on its surface. Each dot represents a galaxy in the universe. As the balloon is blown up all the dots move away from each other. If you observe the speed of a galaxy at some distance away from us, and another galaxy twice that distance, the speed of the farther galaxy traveling away from us is twice the speed of the nearer one. A third galaxy three times the distance is traveling three times the speed of the first galaxy away from us. This ratio goes on until the farthest galaxy visible to us at the edge of our universe is traveling at the speed of light away from us. Once it reaches that speed it is no longer visible to us because the speed of its light traveling toward us is not fast enough to reach us. So all galaxies beyond that outer limit determined by the speed of light are non existent to us, and when adding the expansion of the universe as well, all galaxies now visible to us will reach that point of invisibility.

Not to worry though, because this is a very, very, very long time into the future.
As expansion continues, the universe will become less dense and in fact approach zero density or a state of nothingness. This state is a lot like the "From Nothing" state at the beginning of this paper. Is this again a state of pure mind? A state of infinite genius? And possibly a new beginning for the creative process? One could speculate that this state is again like a field similar to a gravitational or electro-magnetic field. Think of it as a field of infinite spirit and genius. There is enough here to speculate on for both the atheist and theist, or someone like me in the middle, to think about.

NOTES

1.   A writer's license taken here. Recall "A Brief History of Time" by Stephen W. Hawking.

2.   14 billion years times c = 14,000,000,000 x186000 miles/second x 60 seconds/minute x 60 minutes/hour x 24 hours/day x365 days/year = 14,000,000,000 x 186,000 x 31,536,000 = 8.212×10^{22} = 82,120,000,000,000,000,000,000 miles from where we stand looking out to the edge of our universe.

3.   Helium atoms resulted from two hydrogen atoms fused together from the heat and pressure of creation.

4.   This condition is being recreated today by man putting CO_2 and other toxics back into the atmosphere in a matter of years rather than the millions of years it took natural processes to take it out. It is projected that the earth may once again be green without animal life, including us.

5.   Many people today think, from a false evolutionary concept, that we are descended from apes. This is not true. We are all descended from a common ancestor and not from other apes. Who will be

courageous enough to tell them that we are really descended from a rodent sized mammal which in turn originated from a one celled life form.

RECOMMENDED READING

Science News Magazine – Society for Science and the Public.
Discover Magazine – Science for the Curious.
"OH MY GOD" From Doubter to Believer by a Free Thinking Theist.
                    Robert A. Tallarico

RECOMMENDED REFERENCE & STUDY MATERIAL

"EVOLUTION – The Whole Story. Steve Parker.
"ORIGINS" – The Scientific Story of Creation. Jim Baggott.
"VOID" – The Strange Physics of Nothing. James Owen Weatherall.

PHYSICS THAT IS HARD TO UNDERSTAND

QED – The Strange Theory of Light and Matter. Richard P. Feynman.
FIELDS OF COLOR – The Theory that Escaped Einstein. Rodney A. Brooks.

EPILOGUE

END OF LIFE

As I age I find that I more and more think about and search for the reality of the happening that is the end of life. Based on what I know and have experienced thus far in my life I have put together the following hypothesis.

Reading the epilogue of my book "Oh My God!!!" titled "Flight," the thought occurred to me about what happens to people who are dying or have died and are resuscitated. Their experiences vary, but generally involve a bright light, voices soft, gentle and welcoming, sometimes by loved ones, and a feeling of complete euphoria. The mental state persists until the person is either brought back to consciousness or passes away. The last paragraph in the epilogue reads "I continued on my way reflecting on my flight with the swallows wondering what that world would be like for me? If I had a choice, - *when leaving this existence* - I would choose to fly physically unencumbered throughout the unlimited vault of creation, filled with the knowledge of the workings of the universe, and sharing with the Being a common joy that comes from understanding." As I read this I wondered if, when a person dies, their last mental state is what will be their after life based on their own beliefs, experiences and knowledge.

It occurred to me that we who are left behind think in terms of time having a beginning and going on possibly forever. As the result we conceive various ideas of what an afterlife would be like. However we ignore the fact that time is a human device we have created to make sense of observed realities that are time dependent and of creation having a beginning. Time is used by us to establish a sequence for creation and other occurrences "over time." However upon dying, time ends for us. Therefore the last moment we experience is timeless, i.e. it is an eternal now, and as such it doesn't end.

If this is true, then our last state of mind could be our final spiritual state for eternity. This is something to wonder about. In my case could my final experience be as I hoped for in the last paragraph of my piece "Flight."

Also if this is true, than you don't want your last spiritual state to be as is preached to us by various religions, especially the ones that put emphasis on guilt, sin and disobedience of their tenants. How terrible it would be to stand in judgement before the throne for all eternity beating your breast in unending penance because this was your final spiritual state.[1]

Getting back to the major subject, time, which is the crux of this discussion and the eternal now which exists now and existed before the creation of the universe when there was no reality – nothing - there was only the potential for a reality as we know it. There was no time because there was nothing to measure it against.[2]

It was our beginning that the Atheist's Mechanism or the Theist's Creator (they are both the same thing) separated the nothing into positive and negative stuff and time began. This is what led me to believe that a person's state of mind, based on their accumulated knowledge and experience up to that time, is what their eternal now or timeless state will be like. This is why no one (not even the great Houdini) ever comes back to our time to tell us about it. Our time doesn't exist for them. Thus for me, I am hoping my eternal now is as I expressed in my piece "Flight" which is where I am currently in my thinking, or something even more satisfying.

So maybe herein lies the answer to the question "Why". Our advance in wisdom, age and grace is the determining factor in our final state of now and is something to look forward to. Having lived a life filled with purpose, good works and good work, you will experience peace, joy, and unending bliss in your final endless now.

NOTES

1.   This is not meant to degrade the value of religion because religion is essential for those of our kind who do things for rewards or to avoid punishment. Unfortunately religion does come with negative effects which we are witnessing today in the extreme.

There are some people, however, who do things because they are the right thing to do and will result in a better outcome for all of us. This is a humanistic concept which can only work in an idealistic, possibly future, world.

2.   Refer to the beginning of the paper "From Nothing to Now, and Beyond" for a discussion of nothing.

ARTIST'S NARRATIVE

The Painting "genesis"

The painting "genesis" depicts Adam and Eve at the moment of sentencing for disobeying God's directive not to eat the fruit of the tree of knowledge of good and bad.

In the painting we see Adam looking toward God pleading his case while blaming his mate Eve and holding in his hand the source of his condemnation. His penalty: to work endlessly until his death to provide the necessities for his survival. Eve on the other hand, having seduced Adam into sharing the fruit, is looking askance puckishly with the forbidden fruit in her hands partially hidden behind her knowing her penalty is limited to pain on the occasions when giving birth to her children. In the painting she appears to be in the early stages of pregnancy.

We know today that homo became aware when his brain increased to such size to allow him to think abstractly and to project and reflect on images in his mind. Also Eve's pain occurs when she delivers a human child with a large brain of such size as to go beyond her physical ability to deliver without strain and tearing of the birth canal.

The forbidden fruit is the fruit of the vine, mentioned so often in the Old and New Testaments. It has been selected artistically as the forbidden fruit because is it is possible to be fermented naturally on the vine by yeast which can be transmitted airborne or by insects when they bite into the fruit.

Eve, having eaten freely of the fermented grapes, and now with alcoholic clarity, searches for justification in conversation with a snake. He assures her that she is being duped by God and by consuming the forbidden fruit she will be enlightened just like God.

Having been created in a very seductive way to be Adam's mate and to become the mother of the human species, Eve very easily and without hesitation did not hesitate to share this wondrous mind altering fruit with Adam so that he too might become sentient.

The snake, now condemned to crawl on its belly and be derided as being cunning and evil, lives not on the garden floor but in the minds of mankind even as it lived dormant in Eve's mind prior to her consuming the fruit of knowledge of good and evil. It is a quality that we inherit from generation to generation and is a continuous source of our problems. It is a quality that we continue to deal with without much success, considering the condition of the world from our biblical origin in Eve over one hundred thousand years ago, to this day.